"Exercises of Statistical Inference"

SIMONE MALACRIDA

ANALYTICAL INDEX

INTRODUCTION

In this exercise book, some examples of calculations related to statistical inference are carried out.

Furthermore, the main theorems used both in estimation theory and in hypothesis testing are presented.

The study of statistics, in fact, does not stop at the properties of continuous and discrete probability distributions, but expands into inference sectors, applying the statistical concepts of estimation, mean, variance, regression and hypothesis testing when in the presence of particular tests.

In order to understand in more detail what is presented in the resolution of the exercises, the theoretical reference context is recalled in the first chapter.

What is presented in this workbook is generally addressed in advanced statistics courses at university level.

Introduction

I

THEORETICAL OUTLINE

Introduction

Statistical inference falls into two broad areas of interest: estimation theory and hypothesis testing.
At the basis of both areas is sampling understood as the choice of the sample of the statistical population: it can be random, probabilistic, reasoned or convenient.
The sampling methods depend on the probability distribution and on the random variables just described.

Estimation theory

The estimation theory allows to estimate parameters starting from measured data through a deterministic function called **estimator.**
There are various properties that characterize the quality of an estimator including correctness, consistency, efficiency, sufficiency, and completeness.
A **correct estimator** is a function that has an expected value equal to the quantity to be estimated, vice versa it is called biased.
The difference between the expected value of the estimator and that of the sample is called **bias** , if this difference is zero as the

3

sample tends to infinity then the estimator is said to be asymptotically correct.

Given a random variable X of unknown parameter Y, an estimator T(X) is **sufficient** for Y if the conditional probability distribution of X given by T(X) does not depend on Y.

An estimator for the parameter Y is said to be weakly **consistent** if, as the sample size approaches infinity, it converges in probability to the value of Y.

If, on the other hand, it almost certainly converges, then it is said to be consistent in the strong sense.

A sufficient condition for weak consistency is that the estimator is asymptotically correct and that we have at the same time:

$$\lim_{n \to \infty} \mathrm{var}(T_n(x)) = 0$$

We define **Fisher information as** the variance of the logarithmic derivative associated with a given likelihood function (we will define the concept of likelihood shortly).

$$I(\vartheta) = E\left[\left(\frac{\partial}{\partial \vartheta}\ln f(X;\vartheta)\right)^2\right]$$

This quantity is additive for independent random variables.

The Fisher information of a sufficient statistic is the same as that contained in the whole sample.

In the case of multivariate distributions we have:

$$I(\vartheta)_{m,n} = E\left[\frac{\partial}{\partial \vartheta_m}\ln f(X;\vartheta)\frac{\partial}{\partial \vartheta_n}\ln f(X;\vartheta)\right]$$

4

I – Theoretical outline

The Cramer-Rao inequality states that the variance of an unbiased estimator is thus related to the Fisher information:

$$\text{var}\left(\hat{\vartheta}\right) \geq \frac{1}{I(\vartheta)} = \frac{1}{E\left[\left(\frac{\partial}{\partial \vartheta}\ln f(X;\vartheta)\right)^2\right]}$$

In the multivariate case it becomes:

$$\text{cov}_{\vartheta}(T(x)) \geq I(\vartheta)^{-1}$$

The efficiency of an unbiased estimator is defined as follows:

$$e(T) = \frac{1}{I(\vartheta)\,\text{var}(T)}$$

It follows from the Cramer-Rao inequality that the efficiency for an unbiased estimator is less than or equal to 1.

An estimator is said to be efficient if its variance reaches the lower limit of the Cramer-Rao inequality and it is said to be asymptotically efficient if this value is reached as a limit.

The relative efficiency between two estimators is given by:

$$e(T_1, T_2, \vartheta) = \frac{E\left[(T_1 - \vartheta)^2\right]}{E\left[(T_2 - \vartheta)^2\right]}$$

The probability associated with the sample is given by the following probability distribution:

$$P\left(\{x_i\}_{i=1}^n \mid \boldsymbol{\vartheta}\right) = L_D\left(\boldsymbol{\vartheta} \mid \{x_i\}_{i=1}^n\right)$$

To estimate the parameter, the available data x that make up the sample can be used.

The maximum likelihood method searches for the most likely value of this parameter, i.e. that maximizes the probability of having obtained the sample .

In this case the function appearing on the second side of the previous relation is called the likelihood function and the estimator is called maximum likelihood:

$$\tilde{\vartheta} = \arg\max_{\vartheta \in \Theta} L_D(\vartheta \mid x_1,...,x_n)$$

One can choose such estimators to be correct or asymptotically correct.

Furthermore, the maximum likelihood estimator may not necessarily be unique for a given probability distribution.

Given a maximum likelihood estimator for one parameter, then the maximum likelihood estimator for another parameter that functionally depends on the first is given by applying the same function, provided it is bijective.

Maximum likelihood estimators do not reach the lower bound for the variance established by the Cramer-Rao inequality.

The likelihood function is a conditional probability function defined as follows:

$$L(b \mid A) = \alpha P(A \mid B = b)$$

Another method for finding estimators is the so-called **method of moments.**

Using this method, an estimator satisfies the conditions of one or more sample moments.

I – Theoretical outline

It must be said that the maximum likelihood estimators are more efficient than the moments method estimators.

A typical condition of the method of moments is the following:

$$E\left[f(x_i; \vartheta_0)\right] = 0$$

Another estimation method, fundamental for linear regression, is the **least squares method** which allows the identification of trend lines starting from experimental data so that the sum of the squares of the distances between these data and the estimated ones is minimal.

The estimators for the slope and intercept are given by:

$$\hat{\beta}_1 = \frac{s_{XY}}{s_X^2}$$

$$\hat{\beta}_0 = \bar{Y} - \hat{\beta}\bar{X}_1$$

Where we have, in the case of simple linear regression:

$$Y_i = \beta_0 + \beta_1 X_i + u_i$$

While for the multivariate case we have:

$$Y_i = \beta_0 + \beta_1 X_{1i} + \dots + \beta_k X_{ki} + u_i$$

In both cases the statistical error, given by the last parameter u, has zero conditional mean.

Rao-Blackwell estimator is defined as the conditional expected value of an estimator with respect to a sufficient statistic:

$$E\big[\delta(X)\,|\,T(X)\big]$$

The Rao-Blackwell theorem states the standard deviation of a Rao-Blackwell estimator is less than or equal to that of the original estimator:

$$E\big[(\delta_1(X) - \vartheta)^2\big] \le E\big[(\delta(X) - \vartheta)^2\big]$$

So the Rao-Blackwell estimator represents an improvement on the initial estimator.

An estimator is complete if for each measurable function:

$$E\big[g(s(X)\big] = 0 \Rightarrow \forall\,\vartheta,\, P_S(g(s(X)) = 0) = 1$$

The Lehmann-Scheffé theorem states that a correct, complete and sufficient estimator is a minimum variance correct estimator, i.e.:

$$\mathrm{var}(\delta(X_1,...,X_n)) \le \mathrm{var}(\tilde{\delta}(X_1,...,X_n))$$

The Gauss-Markov theorem states that, in a linear regression model having zero expected value error, the best corrected linear estimator is the least squares estimator.

A Bayes estimator is a function that minimizes the expected value of the posterior probability of a function, called loss.

Given a parameter with known prior probability distribution and called L a loss function, then the Bayes risk of the estimator is given by:

$$E_x\{L(\vartheta, \delta)\}$$

The Bayes estimator is the one that minimizes this value.
Under suitable conditions, for a large sample, the Bayes estimator is asymptotically unbiased and converges in distribution to the normal distribution with zero expected value and variance equal to the inverse of the Fisher information, therefore it is also asymptotically efficient.

Hypothesis testing

The second sector of statistical inference is the verification of hypotheses following a statistical test which can be parametric or non-parametric.
A test that can be applied in the presence of a given probability distribution of the data is called parametric, otherwise the test is called non-parametric.
A statistical test involves a statistical error that can be divided into two categories: the first type error is given by rejecting the hypothesis when it is true, the second type is given by accepting the hypothesis when it is false.
This hypothesis is called the null hypothesis or zero hypothesis.

The Neyman-Pearson fundamental lemma states that, given two simple hypotheses, the ratio of likelihood functions that reject the first hypothesis for the second hypothesis is given by:

$$P\big(\Lambda(X) \leq k \mid H_0\big) = \alpha \Rightarrow \Lambda(x) := \frac{L\big(\vartheta_0 \mid x\big)}{L\big(\vartheta_1 \mid x\big)} \leq k$$

And it represents the most powerful hypothesis test.

If this holds for any value of the parameter, then the test is said to be uniformly more powerful.

If we assume the null hypothesis to be true, then the p-value indicates the probability of getting a result equal to the one observed.

It indicates the minimum level of significance for which the null hypothesis is rejected.

The most common parametric tests are given by the Student test where the data distribution is the Student's t or the Fisher test where the data distribution is the Fisher-Snedecor one or the zeta test where the data distribution is a standard normal $N(0,1)$.

In a parametric test it is of fundamental importance to define the confidence interval, ie the range of plausible values for the parameter to be estimated or tested.

The Neyman setting for the confidence interval asserts that it is a set of parameters for which the null hypothesis is accepted.

The confidence level of the interval is given by 1 minus the significance level of the test.

There are many nonparametric tests, let's list some of particular importance.

The binomial test is applied to Bernoulli statistical samples and the probabilities are calculated when the null hypothesis is true.

A test for two dependent samples and to understand the evolution of the situation is the **sign test** which takes into consideration the difference (positive or negative) of the two samples based on the individual parameters.

I – Theoretical outline

Given a binary sequence, the **sequence test is performed to test the independence of the data** .
The number of repeated sequences in a sequence of length N is a normal random variable of expected value and variance given by:

$$\mu = 1 + 2\frac{N^+ N^-}{N}$$

$$\sigma^2 = \frac{(\mu-1)(\mu-2)}{N-1}$$

Where the quotes + and – indicate the positive or negative symbols of the sequence.
Budne's test tests the null hypothesis that two data sets come from two random variables having the same distribution.
The Kolmogorov-Smirnov test tests the shape of sampling distributions and is a nonparametric alternative to the Student test.
The Kruskal-Wallis test tests the medians of different samples for equality.
Pearson's chi-squared test is applied to verifying whether a sample was drawn from a population having a given probability distribution.
If the distribution is binomial then the binomial test can be applied; moreover, if there are at most two samples, the Kolmogorov-Smirnov test can also be applied.
A special case of Pearson's chi-squared test is the **median test** , which tests the null hypothesis that the medians of two samples are equal.
Fisher's exact test is used if the variables are Bernoulli and the samples are small.
The Q test is used to discard or reject statistical data that are not in line with the sample parameters and, therefore, are possible errors.
The Shapiro-Wilk test is used to test the normality of small samples by comparing two estimators for the sample variance.

Regression

A fundamental problem in statistics is regression, i.e. the functional relationship between the measured variables extracted from a potentially infinite sample.
In particular, linear regression is a method of estimating the conditional expected value of a dependent variable Y, once the values of other independent variables X (also called regressors) are known.
The case of simple linear regression is formulated as follows:

$$Y_i = \beta_0 + \beta_1 X_i + u_i$$

The beta values have already been presented as intercept and slope, plus u is the statistical error.
As we have seen, it is possible to estimate these values using the least squares method.
In the case of multiple linear regression, the relationship is as follows:

$$Y_i = \beta_0 + \beta_1 X_{1i} + \dots + \beta_k X_{ki} + u_i$$

The method of least squares allows to find an estimate of the dependent variable which is an orthogonal projection of the vector of observations y on the space generated by the columns of the matrix describing the X independent variables.
The coefficient of determination measures the goodness of fit of the linear regression and is:

$$R^2 = \frac{\sum_i (\hat{y}_i - \bar{y})^2}{\sum_i (y_i - \bar{y})^2}$$

There is also a nonlinear regression that applies to a model of the general form:

$$Y = f(X; \vartheta) + \varepsilon$$

In this case the estimation methods resort to numerical optimization algorithms or linearization processes, introducing an additional error with respect to the statistical error.

Bayesian inference

From Bayes' theorem derives the Bayesian inference approach in which the probabilities are interpreted as levels of confidence for the occurrence of a given event.
In Bayesian statistics, Bayes' theorem takes this form:

$$P(H_0 \mid E) = \frac{P(E \mid H_0) P(H_0)}{P(E)}$$

Where E denotes the observed empirical data, while H_0 it is the null hypothesis, $P(H_0)$ it is called the prior probability, $P(E)$ it is the marginal probability, $P(H_0 \mid E)$ it is the posterior probability, $P(E|H_0)$ it is the likelihood function.

likelihood ratio is called :

$$\Lambda = \frac{L(H_0 \mid E)}{L(notH_0 \mid E)} = \frac{P(E \mid H_0)}{P(E \mid notH_0)}$$

If X is distributed as a binomial random variable having a parameter distributed a priori as a beta then the same parameter distributed a posteriori also follows a beta distribution (obviously with different characteristic parameters).

The same is true if X is distributed as a negative binomial random variable.

If X is distributed as a gamma variable having the second parameter distributed a priori as a gamma then the same parameter distributed posteriorly also follows a gamma.

The same is true if X is distributed as a Poissonian or as a normal one.

II

EXERCISES

Exercise 1

Given a sample set made up of 100 samples, one is chosen at random and the value of 12 is found.

Assume that these values are distributed according to a normal of unknown mean and variance equal to 4.

Propose an unbiased estimator for the mean and find out what its variance is.

Determine the confidence interval for the mean at the 0.95 level.

Determine the same interval without however assuming a normal distribution.

An unbiased estimator is given by the sample mean.

The variance of this estimator tends to zero as the number of samples tends to infinity.

We have:

$$\mathbb{V}ar(\bar{X}) = \mathbb{V}ar\left(\frac{\sum_{i=1}^{n} X_i}{n}\right)$$

$$= \frac{1}{n^2} \sum_{i=1}^{100} \mathbb{V}ar(X_i)$$

$$= \frac{n}{n^2} \mathbb{V}ar(X_i) = \frac{1}{n}\sigma^2$$

And then:

$$\mathbb{V}ar(\bar{X}) \;=\; \frac{1}{100}\sigma^2 = \frac{1}{100} \times 4 = \frac{1}{25}$$

Given the confidence interval, we have:

$$1-\alpha = 0.95 \rightarrow \alpha = 0.05 \rightarrow \alpha/2 = 0.025 \rightarrow$$
$$\Phi(z_{\alpha/2}) = 1-\alpha/2 = 0.975 \rightarrow z_{\alpha/2} = 1.96$$

Since:

$$\bar{X} \sim N(\mu, \sigma^2/n)$$
$$\frac{\bar{X}-\mu}{\sigma/\sqrt{n}} \sim N(0,1)$$

The range is:

$$95\% IC \;=\; [l_1, l_2] = \left[\bar{x} - z_{\alpha/2}\frac{\sigma}{\sqrt{n}}; \bar{x} + z_{\alpha/2}\frac{\sigma}{\sqrt{n}}\right]$$
$$= \left[12 - 1.96\frac{2}{10}; 12 + 1.96\frac{2}{10}\right] = [11.608; 12.392].$$

According to the central limit theorem, this interval remains valid even in the absence of a normal distribution.

Exercise 2

Taking a sample of 25 elements we have:

$$\sum_{i=1}^{25} x_i = 2450.$$

Where each x has a normal distribution with unknown mean and variance equal to 64.
Determine the confidence interval for the mean at the 0.9 level.
Determine the minimum sample size that ensures that the confidence level 0.9 has a length of less than 10.

The estimate of the sample mean is:

$$\bar{x} = \frac{1}{25} \sum_{i=1}^{25} x_i = 2450/25 = 98.$$

$$\mathbb{E}(\bar{X}) = \mu; \quad \mathbb{V}ar(\bar{X}) = \sigma^2/n,$$

$$\bar{X} \sim N(\mu, 64/25)$$

Given that:

$$1-\alpha = 0.90 \rightarrow \alpha = 0.10 \rightarrow \alpha/2 = 0.05 \rightarrow$$

$$\Phi(z_{\alpha/2}) = 1-\alpha/2 = 0.95 \rightarrow z_{\alpha/2} = 1.645.$$

The confidence interval is:

$$
\begin{aligned}
90\%\,IC \;&=\; [l_1, l_2] = \left[\bar{x} - z_{\alpha/2}\frac{\sigma}{\sqrt{n}}; \bar{x} + z_{\alpha/2}\frac{\sigma}{\sqrt{n}}\right] \\
&=\; \left[98 - 1.645\frac{8}{5}; 98 + 1.645\frac{8}{5}\right] = [95.368; 100.632].
\end{aligned}
$$

To enforce that the length of this interval is less than 10, proceed as follows:

$$2 \cdot z_{\alpha/2}\frac{\sigma}{\sqrt{n}} < 10 \Leftrightarrow 2 \cdot z_{0.05}\frac{8}{\sqrt{n}} < 10 \Leftrightarrow$$

$$\Leftrightarrow 2 \cdot 1.645\frac{8}{\sqrt{n}} < 10 \Leftrightarrow 2 \cdot 1.645\frac{8}{10} < \sqrt{n} \Leftrightarrow$$

$$\Leftrightarrow \left(2 \cdot 1.645\frac{8}{10}\right)^2 < n \Rightarrow n > 6.93.$$

And therefore you must have a minimum of 7 samples.

Exercise 3

Given a sample of 60 pieces, defects and weight were found for each of them.
25 pieces are defective and the weights have the following properties:

$$\sum_{i=1}^{60} x_i = 840.$$

$$\sum_{i=1}^{60} x_i^2 = 12300.$$

Propose an unbiased estimator for the mean.
Propose an unbiased estimator for the variance.
Propose a 99% confidence interval in case X is distributed as a normal.
Propose a 99% confidence interval in case X is not distributed as a normal.
Propose a 95% confidence interval for the mean value of Bernoulli's random variable Y which assumes value 1 if the piece is defective and 0 otherwise.

An unbiased estimator for the mean is the sample mean:

$$\bar{x} = \frac{1}{60} \sum_{i=1}^{60} x_i = 840/60 = 14.$$

$$\mathbb{V}ar[\bar{X}] = \frac{1}{n}\sigma^2,$$

This estimator converges in probability.
An unbiased estimator for the variance is the sample variance.

II – Exercises

$$S^2 = \frac{1}{n-1} \sum_{i=1}^{n} (X_i - \bar{X})^2.$$

$$
\begin{aligned}
S^2 &= \frac{1}{n-1} \sum_{i=1}^{n} (x_i - \bar{x})^2 = \frac{1}{n-1} \left[\sum_{i=1}^{n} x_i^2 - n\bar{x}^2 \right] \\
&= \frac{1}{59} \left(\sum_{i=1}^{60} x_i^2 - n\bar{x}^2 \right) = \frac{1}{59} \left(12300 - 60 \cdot 14^2 \right) = 540/59 = 9.15.
\end{aligned}
$$

We have:

$$\frac{\bar{X} - \mu}{S\sqrt{n}} \sim t_{n-1}$$

Given that:

$$
1 - \alpha = 0.99 \rightarrow \alpha = 0.01 \rightarrow \alpha/2 = 0.005 \rightarrow
$$
$$
P(t_{59} \leq t_{59,\alpha/2}) = 1 - \alpha/2 = 0.995
$$
$$
\rightarrow P(t_{59} \geq t_{59,\alpha/2}) = \alpha/2 = 0.005 \rightarrow t_{59,\alpha/2} = 2.66.
$$

The confidence interval is:

$$
\begin{aligned}
99\% IC &= [l_1, l_2] = \left[\bar{x} - t_{59,\alpha/2} \frac{S}{\sqrt{n}}; \bar{x} + t_{59,\alpha/2} \frac{S}{\sqrt{n}} \right] \\
&= \left[14 - 2.66 \frac{3.02}{7.75}; 14 + 2.66 \frac{3.02}{7.75} \right] = [12.96; 15.04].
\end{aligned}
$$

By the central limit theorem, this interval exists even in the case of non-normal distribution.
Finally, for Bernoulli's variable Y we can reason in this way.
The proportion of defective parts is:

20

$$\hat{p} = 25/60 = 0.417.$$

By the central limit theorem, we can say that:

$$\left(\hat{p} \sim N(\pi; \pi(1 - \pi)/n)\right).$$

An estimate of its variance is given by:

$$\hat{\sigma}^2 = \hat{p}(1 - \hat{p})/n.$$

The required confidence interval is:

$$95\%IC \simeq [l_1, l_2] = \left[\hat{p} - z_{\alpha/2}\sqrt{\frac{\hat{p}(1 - \hat{p})}{n}}; \hat{p} + z_{\alpha/2}\sqrt{\frac{\hat{p}(1 - \hat{p})}{n}}\right]$$
$$= [0.417 - 1.96 \cdot 0.063; 0.417 + 1.96 \cdot 0.063] = [0.29; 0.54].$$

Exercise 4

Given a set of samples we have:

$$\sum_{i=1}^{9} x_i = 175.5,$$

$$\sum_{i=1}^{9} x_i^2 = 4222.25.$$

Where the single x's are distributed with a normal of unknown mean and variance.
Propose an unbiased estimator for the mean.

II – Exercises

Propose an unbiased estimator for the variance.
Determine the 99% confidence interval for the mean.

An unbiased estimator for the mean is the sample mean:

$$\bar{X} = \frac{\sum_{i=1}^{n} X_i}{n},$$

In the specific case

$$\bar{x} = \frac{\sum_{i=1}^{n} x_i}{n} = 175.5/9 = 19.5$$

And its variance tends to zero as n increases, in fact:

$$\mathbb{V}ar(\bar{X}) = \sigma^2/n$$

An unbiased estimator for the variance is the sample variance.

$$
\begin{aligned}
S^2 &= \frac{\sum_{i=1}^{n}(x_i - \bar{x})^2}{n-1} = \\
&= \frac{\sum_{i=1}^{n}(x_i)^2 - n(\bar{x})^2}{n-1} = \\
&= \frac{4222.25 - 9(19.5)^2}{8} = \\
&= 800/8 = 100
\end{aligned}
$$

Since:

22

$$\frac{\bar{X} - \mu}{S/\sqrt{n}} \sim t_{n-1}$$

$$1 - \alpha = 0.99 \rightarrow \alpha = 0.01 \rightarrow \alpha/2 = 0.005 \rightarrow$$
$$P(t_8 \leq t_{8,\alpha/2}) = 1 - \alpha/2 = 0.995$$
$$\rightarrow P(t_8 \geq t_{8,\alpha/2}) = \alpha/2 = 0.005 \rightarrow t_{8,\alpha/2} = 3.355,$$

The 99% confidence interval is:

$$
\begin{aligned}
99\%IC &= \left[\bar{x} - t_{n-1;\alpha/2}\frac{S}{\sqrt{n}}; \bar{x} + t_{n-1;\alpha/2}\frac{S}{\sqrt{n}}\right] = \\
&= \left[\bar{x} - t_{8;0.005}\frac{S}{\sqrt{n}}; \bar{x} + t_{8;0.005}\frac{S}{\sqrt{n}}\right] = \\
&= \left[19.5 - 3.355\frac{10}{\sqrt{9}}; 19.5 + 3.355\frac{10}{\sqrt{9}}\right] = [8.317; 30.683].
\end{aligned}
$$

Exercise 5

How does the width of the confidence interval change as the confidence level decreases?

The amplitude decreases, in fact if for example we consider a confidence interval for the mean with known variance:

II – Exercises

$$IC = \left\{ \bar{x} - z_{1-\alpha/2}\sqrt{\frac{\sigma^2}{n}} ; \bar{x} + z_{1-\alpha/2}\sqrt{\frac{\sigma^2}{n}} \right\}$$

Exercise 6

Given a set of 30 samples with variance given by:

$$\sum_{i=1}^{30} (x_i - \bar{x})^2 = 25$$

Assume that they have the normal distribution.
Determine a 90% confidence interval for the variance.

We simply have:

$$
\begin{aligned}
90\% IC &= \left[\frac{(n-1)S^2}{\chi^2_{0.05;29}} ; \frac{(n-1)S^2}{\chi^2_{0.95;29}} \right] = \\
&= \left[\frac{29 \cdot 25}{42.557} ; \frac{29 \cdot 25}{17.708} \right] = (17.036; 40.94)
\end{aligned}
$$

Exercise 7

Determine which of the two estimators S and T is biased and which is the more efficient in the following cases:

II – Exercises

1. $\mathbb{E}(S) = \theta + 2$, $\mathrm{Var}(S) = 3$, $\mathbb{E}(T) = \theta$, $\mathrm{Var}(T) = 8$;

2. $\mathbb{E}(S) = \theta + 2$, $\mathrm{Var}(S) = 5$, $\mathbb{E}(T) = \theta - 1$, $\mathrm{Var}(T) = 6$;

3. $\mathbb{E}(S) = \theta$, $\mathrm{Var}(S) = 6$, $\mathbb{E}(T) = \theta - 1$, $\mathrm{Var}(T) = 4$.

Typically, the mean squared error of an estimator is:

$$EQM[T(X)] = \mathrm{Var}(T(X)) + [\mathbb{E}(T(X)) - \theta]^2.$$

In the first case we have:

$$EQM(S) = (\theta + 2 - \theta)^2 + \mathrm{Var}(S) = 3 + 2^2 = 7,$$
$$EQM(T) = \mathrm{Var}(T) = 8.$$

The estimator S is the most efficient. T is not biased, S is.
In the second case we have:

$$EQM(S) = (\theta + 2 - \theta)^2 + \mathrm{Var}(S) = 5 + 2^2 = 9,$$
$$EQM(T) = (\theta - 1 - \theta)^2 +$$
$$\mathrm{Var}(T) = 6 + 1^2 = 7.$$

Both estimators are biased, T being the more efficient.
In the third case:

$$EQM(S) = \mathrm{Var}(S) = 6,$$
$$EQM(T) = (\theta - 1 - \theta)^2 + \mathrm{Var}(T) = 1^2 + 4 = 5.$$

S is not biased, T is. T is the most efficient.

Exercise 8

Given a population distributed according to a Bernoullian of given parameter, we have an estimator:

$$T(X) = \frac{X_1 + 2X_2 + X_3}{5},$$

Determine if the estimator is correct. If not, calculate the distortion.
Calculate the mean squared error of T.

Since:

$$X \sim Ber(\pi),$$

So:

$$X_1, X_2, X_3 \sim Ber(\pi)$$
$$\mathbb{E}(X_i) = \pi$$
$$Var(X_i) = \pi(1 - \pi).$$

We have:

$$
\begin{aligned}
\mathbb{E}(T(X)) &= \mathbb{E}\left(\frac{X_1 + 2X_2 + X_3}{5}\right) = \\
&= \frac{\mathbb{E}(X_1) + 2\mathbb{E}(X_2) + \mathbb{E}(X_3)}{5} = \frac{\pi + 2\pi + \pi}{5} = \frac{4\pi}{5}.
\end{aligned}
$$

The estimator is incorrect.
Its distortion is:

$$d(T(X)) = 4/5\pi - \pi = -1/5\pi.$$

The variance of the estimator is:

$$\mathbb{V}ar(T(X)) = \mathbb{V}ar\left(\frac{X_1 + 2X_2 + X_3}{5}\right) = \frac{\mathbb{V}ar(X_1) + 4\mathbb{V}ar(X_2) + \mathbb{V}ar(X_3)}{25} =$$

$$= \frac{\pi(1-\pi) + 4\pi(1-\pi) + \pi(1-\pi)}{25} = \frac{6\pi(1-\pi)}{25}.$$

And then the mean squared error is given by:

$$EQM(T(X)) = d(T(X))^2 + \mathbb{V}ar(T(X)) = \frac{1}{25}\pi^2 + \frac{6\pi(1-\pi)}{25}$$

Exercise 9

Given two independent random variables extracted from a normal population of given mean and variance, consider as an estimator of the variance:

$$T_2 = X_1^2 - X_1 X_2$$

Determine if it is correct. If not, calculate the distortion.

Since:

27

$$\mathbb{E}(X^2) = \mathbb{V}ar(X) + [\mathbb{E}(X)]^2 = \sigma^2 + \mu^2.$$

$$\mathbb{E}(X_1 X_2) = \mathbb{E}(X_1)\mathbb{E}(X_2) = \mu \cdot \mu = \mu^2$$

We have:

$$
\begin{aligned}
\mathbb{E}[T_2(X)] &= \mathbb{E}(X_1^2 - X_1 X_2) = \mathbb{E}(X_1^2) - \mathbb{E}(X_1 X_2) \\
&= \sigma^2 + \mu^2 - \mu^2 = \sigma^2.
\end{aligned}
$$

Therefore, the estimator is correct.

Exercise 10

Given a population with unknown mean and variance, consider as an estimator of the mean:

$$T_n = 1/3 X_1 + 1/3 X_2 + \ldots + 1/3 X_n.$$

Find for which values of n the estimator is correct.
How does the variance behave as n increases?

The estimator is correct for n=3.

$$\mathbb{E}(T_n) = \mathbb{E}(1/3 X_1 + 1/3 X_2 + \ldots + 1/3 X_n) = 1/3 n \mathbb{E}(X) = 1/3 n \mu$$

$$1/3 n \mu = \mu \rightarrow 1/3 n = 1$$

The variance is given by:

II – Exercises

$$\mathbb{V}ar(T_n) = Var(1/3X_1 + 1/3X_2 + \ldots + 1/3X_n) = 1/9n\sigma^2$$

So the variance grows with n.

Exercise 11

Given a population with unknown mean and variance, calculate the mean squared error of the following estimator of the mean:

$$T_n = \frac{1}{4}X_1 + \frac{3}{4}\left(\frac{X_2 + \ldots + X_n}{n-1}\right).$$

Given that:

$$\mathbb{E}(T_n) = \mathbb{E}\left[\frac{1}{4}X_1 + \frac{3}{4}\left(\frac{X_2 + \ldots + X_n}{n-1}\right)\right] =$$

$$= \frac{1}{4}\mathbb{E}(X_1) + \frac{3}{4}\mathbb{E}\left(\frac{X_2 + \ldots + X_n}{n-1}\right) =$$

$$= \frac{1}{4}\mu + \frac{3}{4}\mu = \mu.$$

You will have:

$$EQM(T) = \mathbb{V}ar(T) = \frac{1}{16}\mathbb{V}ar(X_1) + \frac{9}{16}\mathbb{V}ar\left(\frac{X_2 + \ldots + X_n}{n-1}\right) =$$

$$= \frac{1}{16}\sigma^2 + \frac{9}{16}\frac{(n-1)}{(n-1)^2}\sigma^2 = \frac{(n+8)}{16(n-1)}\sigma^2$$

Exercise 12

Check whether the following estimator of the mean value of a random variable X is consistent in probability:

$$0.8 \cdot X_1 + \frac{0.2}{n-1} \cdot X_2 + \frac{0.2}{n-1} \cdot X_3 + \cdots + \frac{0.2}{n-1} \cdot X_n$$

An estimator is consistent in probability if:

$$\lim_{n \to \infty} P(|T_n - \theta| < \epsilon) = 1.$$

By Chebyshev's inequality, an estimator is consistent in probability if it is asymptotically unbiased and if its variance tends to zero as n tends to infinity, i.e.:

$$\lim_{n \to \infty} E(T_n - \theta) = 0$$

$$\lim_{n \to \infty} E(T_n - \theta)^2 = \lim_{n \to \infty} Var(T_n) = 0.$$

Let's see if the estimator is unbiased:

$$
\begin{aligned}
E(T_n) &= 0.8 \cdot E(X_1) + \frac{0.2}{n-1} \cdot E(X_2) + \frac{0.2}{n-1} \cdot X_3 + \cdots + \frac{0.2}{n-1} \cdot E(X_n) \\
&= 0.8 \cdot \mu + 0.2 \cdot \frac{n-1}{n-1} \cdot \mu = \mu
\end{aligned}
$$

It's actually not distorted.
Moreover,

$$Var(T_n) = 0.8^2 \cdot Var(X_1) + \frac{0.2^2}{(n-1)^2} \cdot E(X_2) + \frac{0.2^2}{(n-1)^2} \cdot X_3 + \cdots + \frac{0.2^2}{(n-1)^2} \cdot E(X_n) =$$

$$= (0.64 + \frac{0.04}{(n-1)}) \cdot \sigma^2$$

Since this variance does not tend to zero as n tends to infinity, it is not possible to apply what has been said above and therefore the given estimator is not consistent in probability.

Exercise 13

Given a normal population of unknown mean and unit variance, draw a sample of 2 units.
Given the following estimators for the mean:

$$T_{1,2} = \tfrac{2}{3}X_1 + \tfrac{1}{3}X_2$$

$$T_{2,2} = \tfrac{1}{2}X_1 + \tfrac{1}{2}X_2$$

Check if they are correct, determine the most efficient and calculate the efficiency ratio.

We compute the expected values of the estimators:

$$
\begin{aligned}
\mathbb{E}(T_{1,2}) &= \mathbb{E}\left(\frac{2}{3}X_1 + \frac{1}{3}X_2\right) = \\
&= \frac{2}{3}\mathbb{E}(X_1) + \frac{1}{3}\mathbb{E}(X_2) \\
&= \frac{2}{3}\mu + \frac{1}{3}\mu = \mu
\end{aligned}
$$

$$
\begin{aligned}
\mathbb{E}(T_{2,2}) &= \mathbb{E}\left(\frac{1}{2}X_1 + \frac{1}{2}X_2\right) = \\
&= \frac{1}{2}\mathbb{E}(X_1) + \frac{1}{2}\mathbb{E}(X_2) \\
&= \frac{1}{2}\mu + \frac{1}{2}\mu = \mu
\end{aligned}
$$

Both are correct.
The mean squared errors are:

$$
\begin{aligned}
EQM(T_{1,2}) &= \mathbb{V}ar(T_{1,2}) = \mathbb{V}ar\left(\frac{2}{3}X_1 + \frac{1}{3}X_2\right) = \\
&= \frac{4}{9}\mathbb{V}ar(X_1) + \frac{1}{9}\mathbb{V}ar(X_2) \\
&= \frac{4}{9}\sigma^2 + \frac{1}{9}\sigma^2 = \frac{5}{9}\sigma^2,
\end{aligned}
$$

$$
\begin{aligned}
EQM(T_{2,2}) &= \mathbb{V}ar(T_{2,2}) = \mathbb{V}ar\left(\frac{1}{2}X_1 + \frac{1}{2}X_2\right) = \\
&= \frac{1}{4}\mathbb{V}ar(X_1) + \frac{1}{4}\mathbb{V}ar(X_2) \\
&= \frac{1}{4}\sigma^2 + \frac{1}{4}\sigma^2 = \frac{1}{2}\sigma^2.
\end{aligned}
$$

So the more efficient is the second one.
The efficiency ratio is:

$$e = \frac{EQM(T_{1,n})}{EQM(T_{2,n})} = \frac{Var(T_{1,n})}{Var(T_{2,n})} = \frac{5/9\sigma^2}{1/2\sigma^2} = 10/9$$

Exercise 14

Given a population, is a part of it distinguishable with a given characteristic.
Given the estimator for this part of the population:

$$T = \frac{1}{4}X_1 + \frac{3}{4}X_2$$

Determine if it is correct.
Calculate its variance if the population share is 0.3.

The population is distributed according to a Bernoulli law:

$$(X \sim Ber(\pi))$$

Therefore:

$$\mathbb{E}(T) = \mathbb{E}(\frac{1}{4}X_1 + \frac{3}{4}X_2) = \frac{1}{4}\mathbb{E}(X_1) + \frac{3}{4}\mathbb{E}(X_2) = \frac{1}{4}\pi + \frac{3}{4}\pi = \pi$$

So the estimator is correct.
Also, its variance is:

$$\mathbb{V}ar(T) = \frac{1}{16}\mathbb{V}ar(X_1) + \frac{9}{16}\mathbb{V}ar(X_2) = \frac{1}{16}\pi(1 - \pi) + \frac{9}{16}\pi(1 - \pi) = \frac{10}{16}\pi(1 - \pi)$$

Which for 0.3 holds:

$$\mathbb{V}ar(T) = 0.13.$$

Exercise 15

Given a population with normal distribution of unknown mean and variance 256, the mean value is assumed to be 100.
After 64 samples were extracted, an average value of 106 was found.
We want to test the null hypothesis:

$$H_0 : \mu = 100$$

Against the alternative:

$$H_1 : \mu > 100,$$

With a significance value of 0.05.
Determine the acceptance regions and critical region for this test.

The data of the problem are:

$$X \sim N(\mu, 256), \quad \bar{x} = 106, \quad \text{n=64},$$

II – Exercises

$$\alpha = 0.05$$

And the hypothesis is:

$$\begin{cases} H_0 : \mu = 100 \\ H_1 : \mu > 100, \end{cases}$$

We note that:

$$T_n = \frac{\bar{X} - \mu_0}{\sigma/\sqrt{n}} \sim N(0, 1)$$

We reject the null hypothesis and verify the test.
Given that:

$$\alpha = 0.05 \rightarrow \Phi(z_{0.05}) = 1 - 0.05 = 0.95 \rightarrow z_\alpha = 1.645,$$

We have:

$$R.A. : \frac{\bar{x} - \mu_0}{\sigma/\sqrt{n}} \leq 1.645$$

$$R.C. : \frac{\bar{x} - \mu_0}{\sigma/\sqrt{n}} > 1.645.$$

And they are the acceptance and critical regions, respectively.
The observed value of the test statistic under the null hypothesis is:

$$t_n = \frac{\bar{x} - \mu_0}{\sigma/\sqrt{n}} = \frac{106 - 100}{\sqrt{256/64}} = \frac{6}{2} = 3$$

Belonging to the critical region, we reject the null hypothesis.

Exercise 16

Given a population distributed as a Gaussian random variable of unknown mean and variance 3, draw a random sample of 3 items to test the null hypothesis against the alternative:

$$\begin{cases} H_0 : \mu = 2 \\ H_1 : \mu = 1. \end{cases}$$

Given the following critical region:

$$R.C. : \{(x_1, x_2, x_3) : 2x_1 - 2x_2 + x_3 < 1.2\}.$$

Calculate the probability of the first and second type errors.

The test statistic is:

$$T(\mathbf{X}) = 2X_1 - 2X_2 + X_3 \sim N(\mu; 27).$$

Considering that:

II – Exercises

$$\mathbb{E}(T(\mathbf{X})) = 2\mu - 2\mu + \mu = \mu,$$

$$
\begin{aligned}
\mathbb{V}ar(T(\mathbf{X})) &= \mathbb{V}ar(2X_1 - 2X_2 + X_3) \\
&= 4\mathbb{V}ar(X_1) + 4\mathbb{V}ar(X_2) + \mathbb{V}ar(X_3) \\
&= 9\mathbb{V}ar(X) = 27.
\end{aligned}
$$

The probability of an error of the first type is given by:

$$
\begin{aligned}
\alpha &= P(\text{errore di I tipo}) = P(\text{respingere } H_0|H_0) = \\
& \quad P(2X_1 - 2X_2 + X_3 < 1.2|\mu = 2) \\
&= P\left(\frac{2X_1 - 2X_2 + X_3 - \mu}{\sqrt{27}} < \frac{1.2 - \mu}{\sqrt{27}}|\mu = 2\right) \\
&= P\left(Z < \frac{1.2 - 2}{5.2}\right) \\
&= P(Z < -0.15) = \Phi(-0.15) = 1 - \Phi(0.15) = 0.4404.
\end{aligned}
$$

The acceptance region will be:

$$R.A. : \{(x_1, x_2, x_3) : 2x_1 - 2x_2 + x_3 \geq 1.2\}.$$

And so the probability for an error of the second kind is:

$$
\begin{aligned}
\beta &= P(\text{errore di II tipo}) = P(\text{non respingere } H_0|H_1) = \\
& \quad P(2X_1 - 2X_2 + X_3 \geq 1.2|\mu = 1)
\end{aligned}
$$

$$= \quad Pr\left(\frac{2X_1 - 2X_2 + X_3 - \mu}{\sqrt{27}} \geq \frac{1.2 - \mu}{\sqrt{27}} \middle| \mu = 1\right)$$

$$= \quad Pr\left(Z \geq \frac{1.2 - 1}{5.2}\right)$$

$$= \quad Pr(Z \geq 0.038) = 1 - \Phi(0.038) = 0.484.$$

Exercise 17

In a sample of 15 items, the mean value is 28.5 and the variance is 16.

Assuming that the population is distributed according to a normal with mean 25 and unknown variance, verify:

$$Ipotesi \begin{cases} H_0 : \mu \leq 25 \\ H_1 : \mu > 25 \end{cases}$$

With a probability of error of the first type of 0.025.

We have:

$$\hat{\sigma} = S = 4, \quad \bar{x} = 28.50, \quad \text{n} = 15, \quad \alpha = 0.025$$

We use the t-student as it uses an unbiased estimate of the variance:

38

$$T_n = \frac{\bar{X} - \mu_0}{S/\sqrt{n}} \sim t_{n-1}$$

We reject the null hypothesis.
Given that:

$$\alpha = 0.025 \rightarrow P(t_{14} \geq t_{14,\alpha}) = 0.025 \rightarrow t_{14,\alpha} = 2.145.$$

The regions of acceptance and criticism are:

$$R.A. : \frac{\bar{x} - \mu_0}{S/\sqrt{n}} \leq 2.145$$

$$R.C. : \frac{\bar{x} - \mu_0}{S/\sqrt{n}} > 2.145.$$

The observed value of the test statistic under the null hypothesis:

$$t_n = \frac{\bar{x} - \mu_0}{S/\sqrt{n}} = \frac{28.5 - 25}{4/\sqrt{15}} = 3.39$$

It falls into the critical region and therefore it is correct to reject the null hypothesis.

Exercise 18

Given a population distributed according to a normal distribution with unknown mean and variance, extract a sample of 10 items which have mean and variance equal to 12 and 16.
Find a confidence interval for the mean at the 0.95 level.
Test the hypothesis:

$$H_0 : \mu = 13$$

Against the alternative at the 0.05 level.
What if the level is 0.01?

The distribution for the estimator of the mean is one of t-students with n-1 degrees of freedom.
The central limit theorem cannot be applied as there are only 10 reference samples.
The confidence interval sought is:

$$95\% CI = [l_1, l_2] = \left[\bar{x} - t_{n-1;\alpha/2}\frac{S}{\sqrt{n}} ; \bar{x} + t_{n-1;\alpha/2}\frac{S}{\sqrt{n}} \right] =$$

$$= \left[\bar{x} - t_{9;0.025}\frac{S}{\sqrt{n}} ; \bar{x} + t_{9;0.025}\frac{S}{\sqrt{n}} \right] =$$

$$= \left[12 - 2.262\frac{4}{\sqrt{10}} ; 12 + 2.262\frac{4}{\sqrt{10}} \right] = [9.14; 14.86].$$

The test is as follows:

$$\begin{cases} H_0 : \mu = 13 \\ H_1 : \mu \neq 13. \end{cases}$$

II – Exercises

The critical region is:

$$R.C. : = \left| \frac{\bar{x} - \mu_0}{S/\sqrt{n}} \right| > t_{9,0.025}.$$

The critical value is:

$$P(t_{n-1} < -t_{n-1,\alpha/2}) = P(t_{n-1} > t_{n-1,\alpha/2}) = \;$$

$$\alpha/2 = 0.025 \rightarrow t_{n-1,\alpha/2} = 2.262$$

Then the rejection region is given by:

$$R.C. : \begin{cases} \frac{\bar{x} - \mu_0}{S/\sqrt{n}} > 2.262 \\ \frac{\bar{x} - \mu_0}{S/\sqrt{n}} < -2.262 \end{cases}$$

The value of the test statistic is:

$$t_n = \frac{\bar{x} - \mu_0}{S/\sqrt{n}} = \frac{12 - 13}{4/\sqrt{10}} = -0.79$$

Since we do not belong to the critical region, we cannot reject the null hypothesis.
Decreasing the error level to 0.01 does not change anything.

Exercise 19

Given a population with mean 37 and standard deviation 10, we extract 400 samples whose mean is 36. With a significance level of 1%, can we reject the hypothesis that the true mean is 37?

We have:

$$\sigma = 10, \quad \bar{x} = 36, \quad n = 400, \quad \alpha = 0.01$$

$$\begin{cases} H_0 : \mu_0 = 37 \\ H_1 : \mu_0 \neq 37 \end{cases}$$

Given the large number of samples, the central limit theorem can be used.
Therefore:

$$T_n = \frac{\bar{X} - \mu_0}{\sigma/\sqrt{n}} \simeq N(0, 1)$$

The critical region is:

$$R.C. : = \left| \frac{\bar{x} - \mu_0}{\sigma/\sqrt{n}} \right| > z_{\alpha/2},$$

Given that:

$$\alpha = 0.01 \rightarrow \alpha/2 = 0.005 \rightarrow \Phi(z_{0.005}) = 1 - 0.005 = 0.995 \rightarrow z_{\alpha/2} = 2.575.$$

We have:

$$R.C. : \quad = \quad \left| \frac{\bar{x} - \mu_0}{\sigma / \sqrt{n}} \right| > 2.575,$$

The value of the test statistic under the null hypothesis is:

$$t_n = \frac{\bar{x} - \mu_0}{\sigma / \sqrt{n}} = \frac{36 - 37}{10 / \sqrt{400}} = -2$$

Since we do not fall into the critical region, we cannot reject the null hypothesis.

Exercise 20

Given a population, take 400 samples of which 240 are defective. Propose a critical region to test the following hypothesis at a significance level of 1%:

$$\begin{cases} H_0 : \ \pi = 0.5 \\ H_1 : \ \pi < 0.5. \end{cases}$$

We have:

II – Exercises

X=difettosità~ $Ber(\pi)$; $\hat{p} = 240/400 = 0.6$; $\alpha = 0.01$; $n=400$.

Since the number of samples is large, the central limit theorem can be used.
The test statistic is:

$$T_n = \frac{\hat{p} - \pi_0}{\sqrt{\pi_0(1 - \pi_0)/n}} \simeq N(0, 1),$$

The critical region and the acceptance region are:

$$R.C. := \frac{\hat{p} - \pi_0}{\sqrt{\pi_0(1 - \pi_0)/n}} < -z_\alpha$$

$$R.A. := \frac{\hat{p} - \pi_0}{\sqrt{\pi_0(1 - \pi_0)/n}} \geq -z_\alpha$$

Since:

$$\alpha = 0.01 \to \Phi(z_{0.01}) = 1 - 0.01 = 0.99 \to z_\alpha = 2.33, \to -z_\alpha = -2.33$$

The critical region is:

$$R.C. = \frac{\hat{p} - \pi_0}{\sqrt{\pi_0(1 - \pi_0)/n}} < -2.33,$$

Exercise 21

Given a population with mean equal to 3.2, 81 samples are taken such that:

$$\sum_{i=1}^{81} x_i = 234.9 \qquad \sum_{i=1}^{81} x_i^2 = 1001.21.$$

We want to test the null hypothesis:

$$H_0 \; : \; \mu \; = \; 3.2$$

Propose an unbiased estimator for the variance.
Calculate the confidence interval for the mean at the 0.9 level.
Test the hypothesis with significance levels of 0.1 and 0.05.

The unbiased estimator is the sample variance.
We have:

$$\bar{x} = \frac{1}{n} \sum_{i=1}^{81} x_i = 234.9/81 = 2.9.$$

$$S^2 \; = \; \frac{1}{n-1} \left[\sum_{i=1}^{n} X_i^2 - n\bar{X}^2 \right]$$

II – Exercises

$$S^2 = \frac{1}{n-1}\left[\sum_{i=1}^{n} x_i^2 - n\bar{x}^2\right] =$$

$$= \frac{1}{80}\left[1001.21 - 81 \times 2.9^2\right]$$

$$= \frac{1}{80}(1001.21 - 681.21)$$

$$= \frac{1}{80}(320) = 4.$$

Since the number of samples is large, the central limit theorem can be used.
The required confidence interval is:

$$90\%CI = [l_1, l_2] \quad \left[\bar{x} - z_{\alpha/2}\frac{S}{\sqrt{n}}; \bar{x} + z_{\alpha/2}\frac{S}{\sqrt{n}}\right]$$

$$= \left[2.9 - 1.645\frac{2}{9}; 2.9 + 1.645\frac{2}{9}\right] = [2.535; 3.265].$$

The test statistic:

$$\frac{\bar{X} - \mu_0}{S/\sqrt{n}};$$

It is a normal standard.
Therefore,

$$P(T_n < t_n | \mu = 3.2) = P\left(\frac{\bar{X} - \mu_0}{S/\sqrt{n}} < \frac{2.9 - 3.2}{2/9}\right)$$

$$= P(Z < -1.35) = 1 - \Phi(1.35) = 0.09.$$

This result is in the critical region if the significance level is 0.1 but not 0.05.
Then, depending on the levels of significance, the null hypothesis will be rejected or accepted.

Exercise 22

Given a population of unknown mean and variance, 120 samples are extracted with mean equal to 25.3 and variance equal to 13'240.
Test the hypothesis that the mean is zero with a significance level of 1%.
Calculate the actual significance level.
Calculate the probability of error of the second type in the case of an alternative hypothesis with mean = 10.

The critical region is:

$$R.C. = \{z : z \geq z_\alpha\},$$

$$R.C. = \{z : z \geq z_{0.99} = 2.326\}.$$

Since the number of samples is large, we can use the central limit theorem.
Therefore:

$$t_n = \frac{(25.3 - 0)}{\sqrt{13240/120}} = 2.41$$

Falling back into the critical region, the null hypothesis has to be rejected.
The actual significance level is:

$$p = P(\bar{X} \geq 25.3 | \mu = 0) = P\left(\frac{\bar{X} - \mu_0}{S/\sqrt{n}} \geq \frac{25.3 - 0}{\sqrt{13240/120}}\right)$$

$$= P(Z \geq 2.41) = 1 - 0.9920 = 0.008$$

Which is less than 1%, according to what we found earlier. To determine the probability of a type II error, we have:

$$\frac{(\bar{x} - 0)}{\sqrt{13240/120}} \geq 2.326,$$

From which:

$$\bar{x}_c \geq 24.4$$

So the critical region is:

$$R.C. = \{\bar{x} : \bar{x} \geq 24.4\}.$$

This probability is given by:

$$P(\bar{X} \in R.A. | H_1) = P(\bar{X} \leq 24.4 | \mu = 10) = P\left(\frac{\bar{X} - \mu_1}{S/\sqrt{n}} \leq \frac{24.4 - \mu_1}{S/\sqrt{n}}\right) =$$

$$= P\left(Z \leq \frac{24.4 - 10}{10.5}\right) = P(Z \leq 1.37) = 0.915.$$

Exercise 23

Given:

X N. unità	10	20	50	100	150	200
Y Costo unitario	9.4	9.2	9.0	8.5	8.1	7.4

Estimate correlation coefficient and say if it is adequate.
Estimate the parameters of the regression model.

The correlation coefficient is:

$$
r_{XY} = \frac{\sum_{i=1}^{n}(x_i - \bar{x})(y_i - \bar{y})}{\sqrt{\sum_{i=1}^{n}(x_i - \bar{x})^2 \sum_{i=1}^{n}(y_i - \bar{y})^2}} =
$$

$$
= \frac{\sum_{i=1}^{n} x_i y_i - n\bar{x}\bar{y}}{\sqrt{(\sum_{i=1}^{n} x_i^2 - n\bar{x}^2)(\sum_{i=1}^{n} y_i^2 - n\bar{y}^2)}}
$$

In our case we have:

$$
\bar{x} = 530/6 = 88.\bar{3}
$$

$$
\bar{y} = 51.6/6 = 8.6
$$

$$
Dev(X) = \sum_{i=1}^{n}(x_i - \bar{x})^2 = 75500 - 6 \cdot 88.\bar{3}^2 = 28683.33
$$

$$
Dev(Y) = \sum_{i=1}^{n}(y_i - \bar{y})^2 = \ = 446.62 - 6 \cdot 8.6^2 = 2.86
$$

$$
Cod(X,Y) = \sum_{i=1}^{n}(x_i - \bar{x})(y_i - \bar{y}) = 4273 - n \cdot 88.\bar{3} \cdot 8.6 = -285
$$

II – Exercises

$$S_E^2 = \sum_{i=1}^{n} \hat{e}_i^2$$

$$r_{XY} = -285/\sqrt{28683.33 \cdot 2.86} = -0.995$$

The relationship between the variables is not symmetric and therefore the correlation coefficient is inadequate.
It is better to proceed with the regression analysis.
Since:

$$S_E^2 = \sum_{i=1}^{n} \hat{e}_i^2 = S_y^2 - \hat{\beta}^2 S_x^2 = Dev(Y) - \hat{\beta}^2 Dev(X)$$

We have:

$$\hat{\beta} = Cod(X.Y)/Dev(X) = -285/28683.33 = -0.0099$$
$$\hat{\alpha} = \bar{y} - \hat{\beta}\bar{x} = 8.6 - (-0.0099) \cdot 88.3 = 9.4777$$
$$\hat{\sigma}^2 = \frac{Dev(Y) - \hat{\beta}^2 Dev(X)}{n-2} = (2.86 - (-0.0099)^2 \cdot 28683.33)/4 = 0.0071$$

Exercise 24

Data:

II – Exercises

x_i	y_i
-2	2
-5	-3
4	10
5	8
8	20
10	60
-7	-18
12	24

Estimate the regression line and calculate the index of determination.
What is the most outlier?
By removing this value, estimate the new regression line and say whether its index of determination explains more than 90% of the total variability.

We have:

$$\bar{y} = \frac{1}{n}\sum_i y_i = 12.875$$

$$\bar{x} = \frac{1}{n}\sum_i x_i = 3.125$$

$$S_y^2 = \frac{1}{n}\sum_i y_i^2 - \bar{y}^2 = \frac{1}{n}\sum_{i=1}^{n}(y_i - \bar{y})^2 = 468.8594$$

$$S_x^2 = \frac{1}{n}\sum_i x_i^2 - \bar{x}^2 = \frac{1}{n}\sum_{i=1}^{n}(x_i - \bar{x})^2 = 43.60938$$

$$S_{xy} = \frac{1}{n}\sum_i y_i x_i - \bar{y}\bar{x} = \frac{1}{n}\sum_{i=1}^{n}(x_i - \bar{x})(y_i - \bar{y}) = 117.8906$$

Therefore:

II – Exercises

$$\hat{\beta} = \frac{S_{xy}}{S_x^2} =$$

$$= \frac{117.8906}{43.60938} = 2.703331$$

$$\hat{\alpha} = \bar{y} - \hat{\beta}\bar{x} =$$

$$= 12.875 - 2.703331 \cdot 3.125 = 4.427091.$$

And the line is:

$$\hat{y}_i = \hat{\alpha} + \hat{\beta}x_i = 4.427091 + 2.703331x_i$$

We know that:

$$Dev(Y) = n \cdot S_y^2 = 8 \cdot 468.8594 = 3750.8752$$
$$Dev(X) = n \cdot S_x^2 = 8 \cdot 43.60938 = 348.87504$$
$$\sum_{i=1}^{n} \hat{e}_i^2 = Dev(Y) - \beta^2 Dev(X) = 3750.8752 - (2.703331)^2 348.87504 = 1201.296933$$

And so the index of determination is:

$$R^2 = \frac{\sum_{i=1}^{n}(\hat{y}_i - \bar{y}_i)^2}{\sum_{i=1}^{n}(y_i - \bar{y}_i)^2}$$

$$= 1 - \frac{\sum_{i=1}^{n} \hat{e}_i^2}{\sum_{i=1}^{n}(y_i - \bar{y}_i)^2}$$

$$R^2 = 1 - \frac{1201.296933}{3750.8752} = 0.6797$$

II – Exercises

From the observation of the residuals, it can be deduced that the sixth measurement is the most anomalous, having the largest residual.
By removing this measure:

$$\bar{y} = 6.142857$$
$$\bar{x} = 2.142857$$
$$S_y^2 = 173.2653$$
$$S_x^2 = 42.12245$$
$$S_{xy} = 81.83674$$

$$\hat{\beta} = \frac{S_{xy}}{S_x^2} =$$
$$= \frac{\frac{1}{n}\sum_i x_i y_i - \bar{y}\bar{x}}{\frac{1}{n}\sum_i x_i^2 - \bar{x}^2} =$$
$$= \frac{81.83674}{42.12245} = 1.942830$$

$$\hat{\alpha} = \bar{y} - \hat{\beta}\bar{x} =$$
$$= 6.142857 - 1.942830 \cdot 2.142857 = 1.97965.$$

The new correlation coefficient is:

$$r_{XY} = \frac{S_{xy}}{\sqrt{S_x^2 S_y^2}}$$
$$= \frac{81.83674}{\sqrt{42.12245 \cdot 173.2653}} =$$
$$= 0.9579343,$$

For which:

$$R^2 = 0.9579343^2 = 0.9176381 > 0.90.$$

Exercise 25

Data:

$$\sum_{i=1}^{24} x_i = 42.8; \quad \sum_{i=1}^{24} y_i = 12.4;$$

$$\sum_{i=1}^{24} x_i^2 = 81.5; \quad \sum_{i=1}^{24} y_i^2 = 6.72, \quad \sum_{i=1}^{24} x_i y_i = 23.27$$

Calculate the regression line.

We have:

$$\bar{x} = \frac{1}{24} 42.8 = 1.7833; \quad \bar{y} = \frac{1}{24} 12.4 = 0.5167;$$

$$DEV(x) = \sum_{i=1}^{n} x_i^2 - n\bar{x}^2 = 81.5 - 76.3267 = 5.1733$$

And then:

$$\hat{\beta} = \frac{\sum_{i=i}^{n} x_i y_i - n\bar{x}\bar{y}}{\sum_{i=1}^{n} x_i^2 - n\bar{x}^2} = \frac{23.27 - 24 \cdot 1.7833 \cdot 0.5167}{5.1733} = 0.2236$$

$$\sum_{i=1}^{24}(y_i - \hat{\alpha} - \hat{\beta}x_i) = 0 \Rightarrow \hat{\alpha} = \bar{y} - \hat{\beta}\bar{x} = 0.1179$$

II – Exercises

Exercise 26

Data:

x_i	y_i
1.6	10
2	15
3.5	20
3	21
3.2	24
4	30

Calculate the regression line and the index of determination.

We have:

	x_i	y_i	x_i^2	y_i^2	$x_i \cdot y_i$
	1.6	10	2.56	100	16
	2	15	4	225	30
	3.5	20	12.25	400	70
	3	21	9	441	63
	3.2	24	10.24	576	76.8
	4	30	16	900	120
Tot	17.3	120	54.05	2642	375.8
Tot/n	2.88333	20	9.00833	440.333	62.6333

And then:

$$
\begin{aligned}
\hat{\beta} &= \frac{\sum_{i=1}^n (y_i - \bar{y})(x_i - \bar{x})}{\sum_{i=1}^n (x_i - \bar{x})^2} = \frac{n^{-1} \cdot \sum_{i=1}^n x_i y_i - \bar{x}\bar{y}}{n^{-1} \cdot \sum_{i=1}^n x_i^2 - \bar{x}^2} = \\
&= \frac{62.6333 - 2.8833 \cdot 20}{9.0083 - 2.8833^2} = \frac{4.9673}{0.6947} = \\
&= 7.15028
\end{aligned}
$$

$$\hat{\alpha} = \bar{y} - \hat{\beta}\bar{x}$$
$$= 20 - 7.15028 \cdot 2.88333 = -0.6166168.$$

For the index of determination:

$$r_{XY} = \frac{S_{xy}}{\sqrt{S_x^2 \cdot S_y^2}}$$
$$= \frac{4.9667}{\sqrt{0.6947 \cdot (440.33 - 20^2)}} =$$
$$= \frac{4.9667}{\sqrt{0.6947 \cdot 40.333}} = 0.9382.$$

$$R^2 = r_{XY}^2 = 0.9382^2 = 0.8804$$

Exercise 27

Data:

$$\bar{y} = 4$$
$$\bar{x} = 20$$
$$S_y^2 = 2$$
$$S_x^2 = 60$$
$$\frac{1}{n}\sum_i x_i y_i = 68$$

Estimate the parameters of a linear regression model.

II – Exercises

We have:

$$
\begin{aligned}
\hat{\beta} &= \frac{S_{xy}}{S_x^2} = \\
&= \frac{\frac{1}{n}\sum_i x_i y_i - \bar{y}\cdot\bar{x}}{S_x^2} = \\
&= \frac{68 - 20\cdot 4}{60} = \frac{-12}{60} = -0.2 \\
\hat{\alpha} &= \bar{y} - \hat{\beta}\bar{x} = \\
&= 4 + 0.2\cdot 20 = 8.
\end{aligned}
$$

And then:

$$
\begin{aligned}
\hat{y}_{n+1} &= \hat{\alpha} + \hat{\beta}\cdot 22 \\
&= 8 - 0.2\cdot 22 = 3.6
\end{aligned}
$$

9 798215 668986